HOW TO
BUILD A VALUE STREAM MAPPING (VSM)

Step-by-step methodology
Detailed explanations
Examples, tips and tricks

BY MICKAEL RÉQUILLARD

© Mickaël Réquillard, 2020

Éditeur : Mickaël Réquillard, Maine et Loire, France.
ISBN : 9798655440005.
Dépôt légal : Août 2020.
Imprimé à la demande par Amazon.

By the same author :
- How to reduce waste : Discover the 8 'mudas' of any company,
learn how to recognize them to eliminate them !
- How to carry out a 5S project : Step by step methodology, tips
and tricks, toolbox to get started !

INTRODUCTION

Lean management appeared in the 1950s and is now deployed not only in production activities but also in a multitude of functions. Its main objective is to eliminate all process waste in order to improve process performance and meet the right customer needs by allocating company resources wisely.

A multitude of continuous improvement tools are available to achieve this goal, including VSM (Value Stream Mapping). This tool, which may seem complex the first time around, is an excellent way to analyse a process, identify how it works (detailing both physical and information flows) and highlight losses. Once you have acquired the various notions specific to this method (which we will detail in this book), it becomes an excellent means of communication, visual, at any level in the company. It allows to define the improvements and the strategy for the months or even years to come. It allows decision-makers to take a real overview of the situation.

This book will allow you to :
- Understand the terms used in a VSM.
- Understand when to use it and the benefits of this method.
- Know how to read a VSM by understanding its logic, symbols and formalisms.
- Know how to represent a current VSM and design a future VSM.
- Understand how to calculate key indicators.
- Put into practice through several examples.

Enjoy reading!

TABLE OF CONTENTS

1. THE TERMS USED	5
2. WHY MAKE A VSM ?	8
3. HOW TO READ A VSM ?	10
VSM OVERVIEW	10
SYMBOLS USED IN VSM	13
4. REPRESENT THE CURRENT VSM	18
PLOT OF THE CURRENT VSM	18
PLOT OF YOUR FIRST VSM	21
5. DESIGNING FUTURE VSM	26
THE 8 QUESTIONS OF THE FUTURE VSM	26
PRACTICAL APPLICATION	34
6. QUIZZ	41

1.
THE TERMS USED

VSM is a complete method, which brings with it several notions and terms, which it is important to know before starting the exercise. Let's look at this in detail.

Change over time (CO) : the amount of time between the last A-compliant part manufactured and the first new B-compliant part manufactured (at the expected rate).

Contract lead time : The contractual lead time between the customer and the supplier, which is the amount of time between the sales order and the customer receiving the finished product.

Cycle Time (CT) : how often a product is completed in a process, under direct observation. In the case of a manual process, the time that elapses for the operator to complete all the necessary steps before repeating these operations.

FIFO (First In - First Out) : first in, first out, flow logic whereby the materials first entering the company, in the flow, in a process, are the first to be consumed. In contrast, LIFO (Last In - First Out), consumes the most recent material first.

KAIZEN : derived from KAI (change) and ZEN (for better), designates a method of continuous improvement, aiming to bring moderate but permanent improvements, to gain in efficiency, using small financial means but strongly involving the staff. Kaizen blitz allow rapid improvements to be made in a particular area of the company or at company level within a few days, with the participation of the various stakeholders.

KANBAN : label in Japanese, method of stock management attributed to Taiichi Ono (TOYOTA) in the 60s, which allows through a label, a card, an electronic chip, to regulate the production of the upstream station, according to the consumption of the downstream station.

Lead time (LT) : corresponds to the time it takes for a part to pass through the process or value chain, from start to finish.

Little's Law : Lead time = Work in Process / Exit Rate, the Work In Process being the one in progress and the Exit Rate corresponding to flow rate (in pieces per hour for example).

MIFA (Material and Information Flow Analysis) : another name (more explicit !) for VSM, highlighting the 2 components of the method : physical flows AND information flows.

SMED : Single Minute Exchange of Die, a method invented by Shigeo Shingo, which aims to reduce changeover times, in order to increase the number of possible variants over the same production interval.

Takt time : indicates the frequency at which a good must be produced in relation to customer demand. Takt time = available working time / daily customer demand. The available working time corresponds to the working time minus the planned breaks, meal times and downtimes. The daily customer demand is calculated by taking the annual or monthly requirement and dividing it by the number of normal working days in the selected period. The takt time is a kind of metronome of customer demand.

Value-added time / Non value-added time : Value-added time corresponds to the fraction of time that adds value to the product or service being studied. On the other hand, non value-added time does not bring any value, and will therefore be the subject of analysis and actions to reduce it. In VSM, intermediate stock values between stages will be converted to (non-value-added) time.

OEE : Overall Equipment Effectiveness , current performance measurement in the production workshops. The OEE is calculated as follows : OEE = (Number of conform parts / Number of parts theoretically achievable) x 100.

Please note : In addition to machine speeds, this OEE data is important to take into account when analysing the VSM datas. A process step may appear to be a bottleneck when observing the machine speeds, but may no longer be a bottleneck when adding the OEE data. For example, if station A is given for 400 parts/hour with an OEE of 80%, while station B has a rate of 500 parts/hour but a OEE of 60%, the bottleneck will be station B, despite a theoretically higher rate !

Processing time (PT) : corresponds to the time during which the product is worked in production.

Process : referring to the ISO 9001 standard, a process is a set of interactive activities that transforms incoming elements into outgoing elements. This applies to both physical elements and information flows.

Production (or Process) Lead Time (PLT) : corresponds to the time taken by a part to go through all the stages of the value chain, from start to finish. This time also includes raw materials, work in progress and finished product stocks still present in the company, expressed in days.

2.
WHY MAKE A VSM ?

Value Stream Mapping can seem complex at first : a lot of information, a lot of arrows in all directions, numbers everywhere : let's admit it, it's quickly scary for a neophyte !

But when you look closer, after having assimilated the few standard graphic symbols (about ten 'currents'), the fog fades away. The steps are based on the names of the positions as they are named in the field. The numerical values on the VSM are derived from observations or chronos made in the field. It is a pragmatic tool that will give a picture of the value chain at a given moment in time.

But at the same time, let's answer our question : why create a VSM ? There are many answers !

First of all, the VSM presents an up-to-date, synthetic and **visual** picture of the entire value chain, from the supply of raw materials to delivery to the end customer. This tool can really be shared and understood by all. It allows to present the state of the art, the picture of the process as it is, with its constraints, defects, resources and limits.

VSM also allows **each of the players to understand the constraints of upstream/downstream workstations, customer or supplier constraints,** redundant information or possible wastage, quality problems, etc. It will highlight malfunctions, allowing a better understanding of the impact of this or that stage in the achievement of objectives. The interest here is that everyone understands his or her role in the chain, his or her impact in the overall picture.

Since VSM is done 'live', with the actors in the field, it will also **highlight problems** not necessarily known or acknowledged : documentary weaknesses, poorly controlled defects, stocks to be rethought, unbalanced series changes, missing or insufficiently trained resources...or even deviations from written references ! It therefore offers a critical look at the processes observed.

VSM is a powerful tool because it can help to enlighten the line operator on the malfunctions of his workstation or the workstations around him, as well as the company manager, who will then be able to visualize the capability of his value chain to meet customer needs and the possible actions to be taken to achieve this ! Can the tool absorb a 30% increase in orders ? On which item is it important to invest to reduce lead times or improve quality ? **It is a real strategic decision support tool.**

Finally, the primary goal of VSM is **to simplify the process !** VSM is in a way the backbone of value chain improvement, which will be optimized with different actions identified during the development of future mapping (by tackling losses, reviewing physical and information flows, constituting a pull flow instead of a push flow, etc.).

3.
HOW TO READ A VSM ?

VSM, as seen above, is a graphical representation of physical and information flows. Certain standardized symbols are used to represent stocks, flows, process steps, etc. Let's look at this in detail.

VSM OVERVIEW

Let's take a look at what a VSM looks like, through the example opposite.

In this simplified example, we will distinguish 5 'groups' of essential information that are present in all VSM :

1. **Customer-specific data,** in zone A in the above example (with the factory symbol in the upper right-hand corner of the VSM). This is where the quantified customer demand (x pieces per month/per day), the possible delivery packaging unit to be taken into account, the customer's work organization (in 1, 2 or 3 shifts for example) and the frequency of supply to the customer (symbolized by the truck) will appear.

2. **The data of the physical flow,** in zone B in the example. You will find here the different stages of the process (the 'boxes' Item A, B, C, D, E, shipments), their characteristics (CT cycle time, CO changeover time, available time, OEE, number of operators, or any other important information in the analysis), the inter-operation stock information (symbolized by the

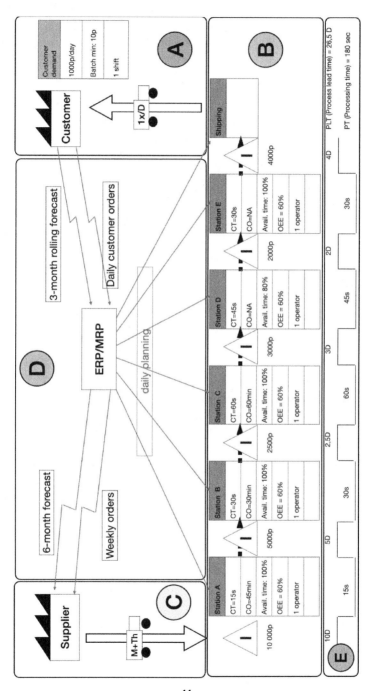

triangle with an I) and the type of flow between each stage (push flow, FIFO, pull flow, etc.).

3. **The supplier data,** in zone C on the example (with the factory symbol in the upper left corner of the VSM). The frequency of supply to the factory is indicated in the truck going to the B zone.

4. **The data of the information flow,** in zone D in the example. This part indicates all the material (right arrows) or electronic (lightning arrows) information that passes through the process : input of customer or commercial information, distribution of schedules to production, transmission of orders or forecasts to suppliers, etc.

5. **The timeline,** in the lower part, in zone E in the example. It will indicate the value-added times (at the bottom of the slots) and sum them to obtain the processing time (here 180 seconds). It will also transcribe (at the top of the slots) the intermediate stocks converted into time (which corresponds to the average 'stay' time in the process, bringing this stock to the number of days of coverage with respect to the customer demand). Thus in our example, the customer demand is 1000 pieces per day. The stock upstream of the station E of 2000 pieces thus constitutes 2 days of stock, which we find indicated on the corresponding slot. The sum of the times of the high slots will indicate the Process Lead Time (PLT) : 26.5 days in our example.

This timeline also shows the reactivity rate = 100 x Processing Time / Process Lead Time.

In our example, the reactivity rate would therefore be 100 x 180 / (26.5 D x 24 H x 60 min x 60 sec) = 0.007%, which is particularly bad ! That said, the reactivity rate will rarely exceed 1% in the majority of cases, as components spend most of their lives moving around or waiting ! In other words, less than 1% of the component's dwell time in the process is value-added !

SYMBOLS USED IN VSM

We've just met face to face with a VSM just now. Now let's look at the symbols used to make them.

We will classify these symbols into three broad categories :
1. Symbols used for the material flow
2. Symbols used for the information flow
3. Symbols used for the time graph

1. <u>Symbols used for the material flow</u>

Station A
CT=15s
CO=45min
Avail. Time : 100%
OEE = 60%
1 operator

Process step box. This box will include :
- The task performed in the header
- Cycle time (CT)
- Changeover time (CO)
- Available time
- OEE (or performance measure)
- Number of operators
- Any other information useful for analysis (batch sizes, quality problems, recurring breakdowns, etc.).

 Symbol used for both the customer (top right) and the supplier (top left).

 Storage symbol.

2500p

 Symbol of a safety stock.

 Symbol of a supermarket : its supply from the upstream station depends on what the downstream station consumes.

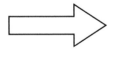 Symbol for the movement of finished products to the customer (i.e. between the process and the customer and between suppliers and the process).

 Symbol used to indicate a pushed flow.

Symbol used to indicate a pulled flow.

 Symbol indicating a FIFO sequence.

 Symbol for departing goods.

 Symbol to designate an operator.

 Symbol for a material handler.

2. Symbols used for the information flow

 Material information flow.

 Electronic information flow.

 Production Kanban.

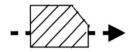 Kanban in pull flow.

 Production smoothing.

 Planning visual.

 Kaizen.

 Kanban box.

3. <u>Symbols used for the time graph</u>

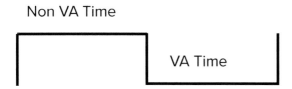

As you can see, most of the symbols are self-explanatory, making it easier to read the VSM. Now that you have the translator, let's get down to business !

4.

REPRESENT THE CURRENT VSM

PLOT OF THE CURRENT VSM

In order to represent the current VSM, we will first have to **identify what we will want to observe**. This first step is crucial. You will have to choose the 'thing' you observe, which you will have to follow through the whole process, from raw material to shipment.

To do this, you will have to choose, for example, a flow on which malfunctions are recurrent. Or the flow in which a new market will generate a strong increase in activity. Or else you will have to create a matrix of choices, to prioritize a production or order flow, either by the volume generated or the turnover. The aim of this first step is therefore to clearly define the contours of the chosen perimeter, what will be observed and what will not. You will thus gain in efficiency.

Once this step has been completed, you will **form the team** that will work on VSM. It will obviously be important to include the stakeholders of the major steps in the process, those who know the flow. People from outside the department, or from outside the company (such as subcontractors) can also help you bring a different perspective to the analysis, other constraints but perhaps also other solutions that you might not have imagined ! Having a quality/process-oriented person in the team can also be a plus to help formalize the different steps of the process (incoming, outgoing, customers and suppliers, nature of the tasks in the process).

Then, together with the team, you will **define the main steps of the process** within the scope initially defined. For these steps, make sure you identify the customers and suppliers, what goes in and what goes out. This will already shed some light on the overall picture.

Then, you **go on the field**, with a big sheet of paper, a pencil, an eraser, a smartphone (which will do calculator, camera and stopwatch) and forward ! You will familiarize yourself with the steps identified above, the goal being to take your bearings, recognize the contours of the perimeter, make a first observation of the flows to understand the overall picture. I advise you to start from the end (expeditions) and to go back up the flow, which allows you to target the flow to be studied from the beginning of the diagnosis, without getting lost in the crossings. Then repeat this exercise by going into details. Remember to observe the physical flows and also to ask questions about the information flows (where does the production schedule arrive ? Where are the orders transmitted ? How often and by whom ? etc.).

Tip : to fill in the process steps and their information and to save time for the plotted representation, I suggest you use repositionable notes prepared in advance (with the names of the steps and the necessary information fields). This way you will not waste time drawing your data boxes and can easily move them around on your paper during the observation.

When making these observations, take measurements in the field, count work in process, measure cycle times : take the information, do not take for granted what is in the routings or on charts. Remember what we saw at the beginning : the aim of the exercise is also to compare the data written in procedures or spreadsheets with the reality of the field, even if it means creating some surprises ;-).

As far as formalism is concerned, I think it's best to display the VSM on a large sheet of paper (like a plotter sheet or a large kraft sheet). Indicate in order :

- The customer part (quantify the daily, weekly, monthly demand, possible constraints such as lot sizes or frequency of supply requested).
- The main steps collected and the information from the associated data boxes on repositionable notes.
- The physical flow (indicating the sequence of tasks), the type of flow between steps (push, pull or FIFO), the intermediate stocks in the flow, up to the raw material.
- The supplier data.
- The flow of information that goes between each part of the flow.
- The time line, fed by the cycle times and the intermediate stocks converted into time.
- Calculate the reactivity rate of the studied flow. Don't leave right away !

Tip : some steps of the process (or even some flows) are sometimes carried out in parallel. In this case, you will need to represent the step boxes one above the other in the VSM. We will only keep in the timeline the longer of the two.

You will probably have to go back a few times to take data, check an information, check a work in progress, etc. The goal is to give a visual picture that is as true to reality as possible on the day of the observation.

Don't succumb to the temptation to put this VSM on computer, it would be a waste of time. Even if tools nowadays make it possible to do so, it is extremely time-consuming. Concentrate your time and energy on making the current VSM picture accurate and visual so that communication with the team and decision-makers is made easy, that is its vocation !

PLOT OF YOUR FIRST VSM

Let us now make a simplified practical example to understand the different principles seen previously. Prepare a blank sheet of paper, a pencil, an eraser, a calculator. Let's go !

Welcome to the SECUTOP company, which has a packaging line to produce car safety kits, including 1 one-size yellow safety vest, a warning triangle, a fire extinguisher, a survival kit and a pair of gloves, all packaged in a zip-up kit. They are then packed in cartons of 5.

The packaging line, U-shaped, is organized as follows :
– Station 1 : Shaping the kit, adding and strapping the extinguisher. CT 9 sec. Availability of the station at 100%. OEE at 80%.
– Station 2 : Addition of the warning triangle and the yellow safety vest. CT 4 sec. Availability of the station at 100%. OEE at 80%.
– Station 3 : addition of gloves and survival kit, closure of the kit. CT 6 sec. Availability of the station at 100%. OEE at 80%.
– Station 4 : Checking of completeness by weighing, placing a sticker and packing in boxes of 5 and palletizing. CT 7 sec. Availability of the station at 100%. OEE at 80%.

Customer demand is for 4000 survival kits per day.
The team works 7 hours a day, on a shift.
Orders leave every day to the customer, deliveries of raw materials take place on Monday and Thursday every week.
Customer orders are received by email every day.

The forecasts of the salesmen are given on 3 months rolling each month by e-mail.

A 6-month forecast is sent to suppliers by e-mail each month.
E-mail orders are sent daily from the information system to the suppliers.

The paper production schedule is sent to the first workstation every day, as well as the paper shipment schedule.

The following stocks were observed during the VSM in the field.
25,000 units in warehouse stock.
2500 units between station 1 and station 2.
4000 units between station 2 and station 3.
4500 units between station 3 and station 4.
10,000 units in finished products.

3 operators work in the zone.

Represent the VSM of this line, freehand on a blank sheet of paper with these datas (before looking at the next page ;-)).

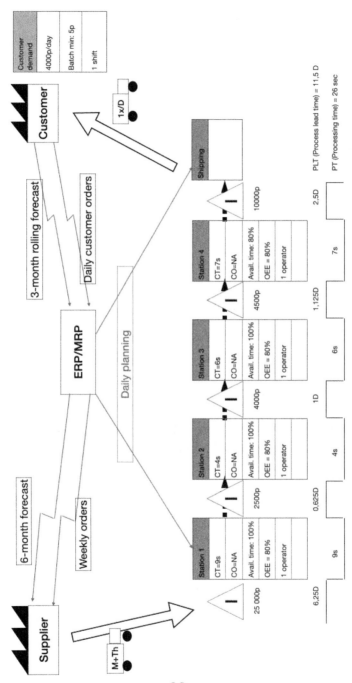

Have you traced your VSM, indicated customer, material flow, supplier, information flow and timeline ? Congratulations !

Now let's ask ourselves a few more questions with this VSM in front of us :

1. *What is the takt time ?*
2. *What is the current cycle time, with 3 operators ?*
3. *What is the processing time ?*
4. *What is the process lead time ?*
5. *What would be the cycle time with 4 operators ?*
6. *What is the reactivity rate of this process with 3 operators and with 4 operators ?*
7. *Can we respond to customer demand with 3 operators ?*

Take a paper, a pencil, a calculator and make these measurements before going to see below ;-)

1. takt time = daily available work time/customer request. Here, work takes place on a single shift of 7 hours, with a break of 2 times 10 minutes. The customer demand is 4000 kits/day. So here Takt time = (7h x 60min x 60sec - 20min x 60sec)/4000 = 6 sec.

2. To calculate the cycle time, the sum of all step cycle times is calculated here and divided by the number of operators. Cycle time with 3 operators = (9 sec + 4 sec + 6 sec + 7 sec)/3 operators = 8.7 sec.

3. The processing time will be the sum of the value-added times. So Processing time = 9 + 4 + 6 + 7 = 26 sec.

4. The process lead time is calculated by summing the values at the top of the slots in the time line (corresponding to the found intermediate stock divided by the customer request). Process lead time = (25000/4000) + (2500/4000) + (4000/4000) +(4500/4000) + (10000/4000) = 11.5 D.

Note : to be completely accurate in the calculation of the process lead time (PLT), you would have to sum the values of the high slots + the values of the low slots, but 26s compared to 11.5D will be almost invisible in the final result, which is why you will very often find the calculation of the process lead time with only the sum of the intermediate stocks.

5. Cycle time with 4 operators = (9 + 4 + 6 + 7) / 4 = 6.5 sec.

6. Reactivity rate with 3 operators = 100 x Processing Time / Process lead time = 100 x 26 sec / (11.5 D x 7 h x 60 min x 60 sec) = 0.008%. Reactivity rate with 4 operators = ...same as with 3 operators ! The intermediate stocks remain the same and the sum of the value-added times also remains the same. Only the cycle time is improved with 4 operators, but does not influence the formula for calculating the reactivity rate.

7. As seen previously, the Takt time is a kind of metronome of the customer's request. Thus, by schematizing, the customer 'consumes' in our example a kit every 6 sec. The cycle time with 3 operators is 8.7 sec. So in this schematic, we will not be able to meet the customer's demand. By adding an operator, we had calculated a cycle time of 6.5 sec. By configuring the cell with 4 people, and by studying optimizations, we will be able to meet the need (it will be necessary to go under 6 sec).

This example allowed us to better understand important concepts in calculations useful in VSM, how to see them, how to calculate them, and how to interpret the results.

5.
DESIGNING FUTURE VSM

We have seen in the previous chapters the terms used, the symbols, the formalism of VSM and how to calculate the key notions of current VSM. Very good !

Note : Stay alert on one point. Think GLOBAL for the proposed improvements, a local improvement without taking into account the upstream/downstream constraints will not necessarily bring the expected effect !

THE 8 QUESTIONS OF THE FUTURE VSM

The second big step will be to criticize (in the good sense of the word !) the current VSM in order to define, design, the future VSM. We will then ask ourselves 8 questions (proposed in the book by Mike Rother and John Shook, *Learning to see*):
1. What is the reference Takt Time ?
2. Do we need a stock of finished products ?
3. Where can we set up a continuous flow ?
4. Where can we foresee Kanban supermarkets ?
5. At what stage should the production order be injected ?
6. How do we schedule the different variants on the regulator station ?
7. What is the picking rate at the regulator station ?
8. Which Kaizen improvement projects are to be planned ?

Let's look at each of these questions in detail.

1. What is the reference Takt Time ?

This Takt Time is calculated by dividing the available working time by the customer demand over the same period. So you will have to remove breaks, lunch breaks, scheduled shift points, everything that is somehow stuck in the daily agenda and is not production time.

In the previous example, if the team works on a single shift of 7 hours during the day, with 2 breaks of 10 minutes (morning and afternoon), the available time will be in our previous example of : (7h x 60min x 60sec) - (2 x 10min x 60sec) = 25200 - 1200 = 24000 sec.

With the daily customer demand of 4000 pieces, we had therefore found a reference Takt Time of 6sec.

This data is indeed a reflection of the customer's demand for a smoother consumption. Unfortunately, the demand will not always be so linear (it would be too simple !). This data will therefore shed light on the capacity of the process to meet the demand or on the actions to be taken globally or on a specific point of the process.

In our previous example, the TT is 6 sec, and the cycle time was 6 sec at 4 operators, after some improvements. We can say that everything is fine. But what if the demand has a seasonality ? What if there are technical problems on the line ? Can the bottleneck station be doubled ? Will it be possible to switch to two teams or three teams, with sufficient staff and with sufficient skills ?

Beyond pure calculation, these are also the questions that need to be asked. I told you it was also a strategic tool ;-)

2. Do we need a stock of finished products ?

Two philosophies of production 'types' will generally emerge :
— Production to order : the customer order is produced specifically for the customer's need (which may be specific to him) and goes directly to shipping, without any storage stage.
— Production to stock : the production will feed a storage warehouse, from which the shipments will go according to the customer's demand.

Both schemes have their advantages and disadvantages:
— Production to order will have the advantage of limiting stocks and therefore fixed assets (either in value or in m3). It is the sales order that will trigger supplies and production. But this solution will also have the disadvantage of a longer reactivity time requiring very good sales forecast reliability (to be able to identify the necessary resources at the right time) and excellent supplier reactivity ! It will adapt well to a regular, stable demand (all things considered).
— Stock production will have the opposite advantages and disadvantages of the previous model. The customer's order will pick up from the stocks, which will then generate new production if necessary (via the implementation of production kanbans, for example, in the form of cards or integrated into the ERP/MRP). This solution will require larger fixed assets (raw material stock, finished goods, storage volumes). On the other hand, it will allow a quicker response to a customer request, or to better cope with seasonality, on more 'standard' types of materials.

Here again, the choice will be essentially strategic. Will tomorrow's production have to respond to a 24-hour dispatch following customer demand in order to differentiate itself from the competition ? Will it be 'standard' products or customized products ? What are the constraints in terms of surface area or storage volumes ?

One of the possible actions can be to determine if certain sub-assemblies or semi-finished products can be made up upstream, in order to save time when taking the order.

3. Where can we set up a continuous flow ?

When you study your VSM you will find that some cycle times are not properly balanced and some may exceed the Takt Time.

Let's illustrate this with the example below :

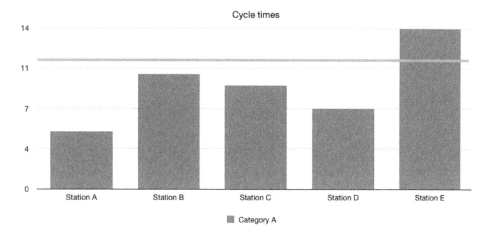

	Category A			
Station A	5			
Station B	10			
Station C	9			
Station D	7			
Station E	14			

Here we have 5 workstations to follow, with distinct cycle times (ranging from 5 to 14 seconds).

I materialized the Takt time by the horizontal line, at 12s.

One workstation will be problematic : E, with a cycle time greater than the takt time, it will 'brake' the flow.

In the same way, we will logically observe waiting times in the process (important at stations A and D here) and significant intermediate stocks in front of stations B and E.

What improvements could be made ?

Summing up all cycle times, we arrive at 45 seconds for 5 stations, an average of 9 seconds per station, well below the required takt time.

By redistributing certain tasks in order to better balance the cycle times, the takt time could be kept to 4 people (45 seconds on 4 shifts = 11.25 seconds).

To do this, several improvements will be made:

- Identifying the most judicious division of tasks in order to obtain the closest possible cycle times between stations.
- Identify which steps of the process could be improved (product design, workstation design, optimised production changes, etc.) to meet the takt time target.
- To work on the versatility of the actors of the process, in order to be able to carry out all the tasks in a fluid way and to compensate for possible absences.

4. Where can we foresee Kanban supermarkets ?

Continuous flow cannot always be set up. In this case, Kanban supermarkets are used.
What does this consist of ?
A kanban will make it possible to regulate upstream production according to downstream consumption. Don't frown, let's take an example ;-).
The kanban should be understood as an 'empty/full tank'.

Let's take the example of a fast food. A limited number of burgers are available for sale. If the customer orders a burger A, the order picker will take it from the 'shop' between the checkout and the kitchen (picking from a predetermined stock). He will make a Kanban call (in the form of an easel or card) to the kitchen, which must then replenish the stock of burger A (production kanban).
Thus, by this simple means, upstream production (in the kitchen) is regulated according to downstream consumption (at the checkout). And so the customer is served faster, stocks are lower (no burgers prepared long in advance and to be reheated), and there is less stock than if all the burgers planned to be sold for the day had been prepared !
This example could be symbolized as follows :

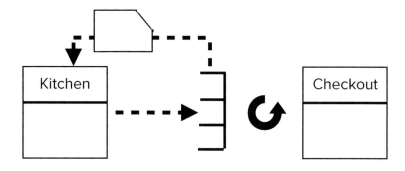

5. At what stage should the production order be injected ?

As you may have noticed, stage is here in the singular ! The purpose of the question here is to define which is the regulating station, i.e. to define at which (single) stage of the process we will make the production schedule available. Indeed, putting several production schedules in a workshop will often imply several losses, such as a difficult update of the schedules, or a delicate management of the upstream / downstream process.

Generally, we will have two scenarios :
- The manufacturing order is given at the first step of the process, with FIFO or continuous flow management of the following steps.
- The production order is given at the last step of the process, if the previous steps are linked by a pull flow (and therefore intermediate supermarkets).

In other words, production planning (and therefore production orders) will be injected at the first stage after the pull flow.

In the example of fast food that we mentioned earlier, we are in the second case. The pull flow will be 'launched' with the production order from the last step of the process (customer order), materialized by the kanban.

6. How do we schedule the different variants on the regulator station ?

This is a difficult but crucial question. As a producer, it is simpler to launch large manufacturing orders, in large quantities, to limit changes, set-up times, production stoppages, etc. On the other hand, if one thinks in terms of the company as a whole, this solution also has many disadvantages. It will generate overstocks, which will immobilise the industrial tool, logistics areas, resources,

32

etc. It will also extend lead times. In fact, running three series of products per day in production will give additional constraints but will reduce stocks of finished products and also be much more responsive to customer demand than by choosing to run one series per day.

Passing several changes during the day can speed up the flow. However, there will be improvements to be made to achieve this goal, among others:
- Optimize changeover times with several SMED.
- Reduce the size of the manufacturing orders.
- Well define the production rhythm to be as efficient as possible.

7. What is the picking rate at the regulator station ?

The aim of this question will be to determine the right batch size to be consumed at the regulating station, in order to deduce the cutting time to operate, respecting the balance between the products injected into the process and the finished products discharged in order to maintain a tight flow.

8. Which Kaizen improvement projects are to be planned ?

The seven previous questions helped to shed light on the process studied and to define several changes to be made (cycle times that are unbalanced or higher than Takt Time, the need to set up Kanban supermarkets, points that need to be reviewed in order to set up a continuous flow, etc.).

All these deviation points will be symbolized on the current VSM by KAIZEN bubbles.

The purpose of this last question is therefore to identify the gaps on the current VSM (which will have been partly done with the previous questions), to determine the local and global actions to be carried out on the process (SMED projects, balancing of workstations, revision of lot sizes, etc.), indicating the carriers and deadlines.

PRACTICAL APPLICATION

Let's go back to the SECUTOP VSM we studied on page 23. Let's ask ourselves the 8 questions that define the future VSM.

1. What is the reference Takt Time ?

That first issue had been addressed. Takt time = available work time/daily customer request. Here, work takes place on a single shift of 7 hours, with a break of 2 x 10 minutes. The customer request is 4000 kits/day. So here Takt time = (7h x 60min x 60sec - 20min x 60sec)/4000 = 6 sec.

2. Do we need a stock of finished products ?

In this VSM, the customer is delivered daily (truck to the right of the VSM, daily departure). It is therefore possible to consider a grouping of the order to shipments, without the need for a stock of finished products.

3. Where can we set up a continuous flow ?

The cycle times are quite unbalanced between workstations :
Station 1 : CT = 9 sec
Station 2 : CT = 4 sec
Station 3 : CT = 6 sec
Station 4 : CT = 7 sec
Graphically represented, it looks like this :

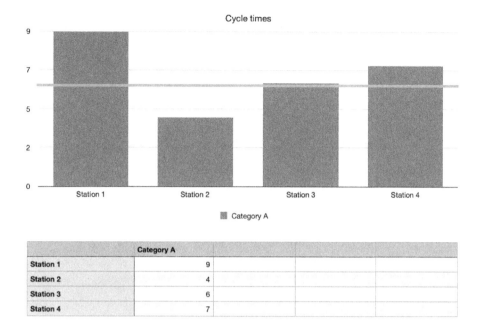

	Category A			
Station 1	9			
Station 2	4			
Station 3	6			
Station 4	7			

With 4 operators, we had a cycle time of 6.5 sec, higher than the 6 sec takt time (horizontal line on the graph).

The nature of the stations and operations must therefore be re-examined in order to balance the times per station but also to keep the customer takt time.

For station 1, which is the most critical, the shaping of the cover will be facilitated by a template, which will speed up the operation. The strapping design will be revised with elastic straps.

Station 2 will add the sticker initially applied in step 4 and the gloves.

Station 3 will check the weight of the completed kit directly on his station.

Station 4 will do all the packaging, palletizing and evacuation of the finished products.

This rebalancing will have allowed a better balance of tasks and consequently create a continuous flow.

4. Where can we foresee Kanban supermarkets ?

Given that we have a continuous flow in place, there is no need here to put a Kanban supermarket in the flow. However, a kanban to supply the raw materials to each of the items can be considered, so that the elements required to make up the kit are always present at the station. Indeed, we have seen that the cycle time, even with rebalancing, was extremely close to the takt time. It will therefore be necessary to ensure that everything that revolves around this process (including the supply of the stations) is optimised.

5. At what stage should the production order be injected ?

In our example, we'll have a FIFO flow instead. In fact, there will be no intermediate supermarkets here. So the production order will be put at station 1, in process input.

6. How do we schedule the different variants on the regulator station ?

In the example taken here, the kits are all identical, no variations in the compositions, so the question is not applicable here.

7. What is the picking rate at the regulator station ?

At the regulating station (station 1 here), the picking could be done for example by 100 carton boxes (500 kits), one pallet. We would not make an order to each kit, imagine the reams of paper !

8. Which Kaizen improvement projects are to be planned ?

Let's have an overview, and identify the projects to be carried out.

We have the right rhythm of dispatch, with daily pick-up, with no additional stock in the supermarket.

We have to balance the stations (as seen in question 3) to put them in flow to arrive around 6 seconds per station, this by reviewing the times and revising the ways of doing things or the design of the product.

The continuous flow and the challenging takt time will require us to review the supply modes of the stations, to avoid any flow stop.

All the intermediate stocks present in the initial VSM will no longer need to be in the future VSM, there will eventually be up to 2 kits in inter-stations.

The stock at the entrance of the flow also needs to be reviewed. There is today the equivalent of 6 days of orders in stock (25000 pieces before item 1 for 4000 kits in daily departure). Deliveries take place twice a week. Even if we secure with the eventuality of a missed truck in reception, we can reasonably imagine to reduce this stock from 6 days to 4 days.

The flow of information will be simplified by transmitting the production schedule only to station 1 (regulating station).

Do the forecasts transmitted to the supplier need to have an amplitude at 6 months, whereas the commercial forecasts do not exceed 3 sliding months ? It is better to secure this information over 3 months.

Orders sent on a weekly basis can be sent on a daily basis.

Let's now look at the graphical representation of these remarks (page 38), before plotting the future VSM (page 39).

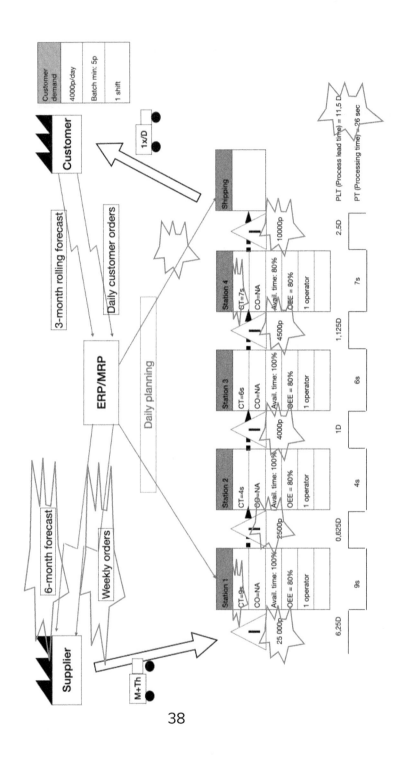

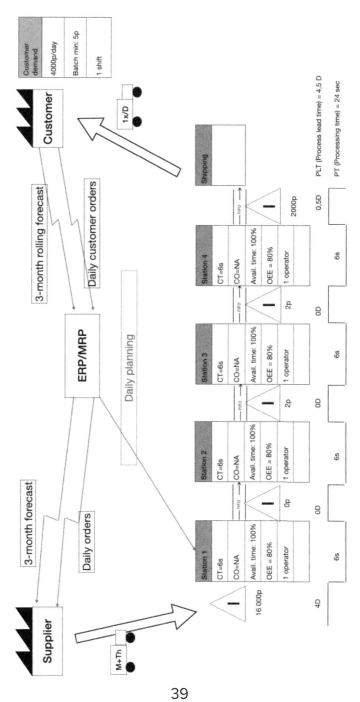

Once all the recommendations have been made, let's take a look at the timeline :

The lead time has gone from 11.5 days to 4.5 days.

The processing time has been reduced from 26s to 24s.

The reactivity rate was in the initial VSM of 0.008%.

This same rate is now : 100 x Processing Time / Process lead time = 100 x 24 sec / (4.5 D x 7 h x 60 min x 60 sec) = 0.02%. This is a 2.6-fold improvement !

This simple example allowed us to familiarize ourselves with the important notions of a VSM : how to read it, how to calculate it, the 8 questions to ask to identify potential improvements to finally draw the future VSM.

At the end of the future mapping, identify the actions to be taken, with their carriers and realistic timeframes to succeed in reaching the objective, as well as an evaluation of the estimated gains, which can help you prioritize the actions (and sell the changes ;-)).

6.

QUIZZ

I suggest that you validate together with the 10 questions of this quiz that the important points related to VSM are acquired.

1. Where are the value-added times represented ?
   A. In the lower part of the time line.
   B. In the upper part of the time line.
   C. In the upper right client part.
   D. In the suppliers' part at the top left.

2. What does Takt Time stand for :
   A. To the time it takes to go through the whole process.
   B. At the customer's request in number of pieces per hour.
   C. The frequency with which a good must be produced in relation to the customer's request.
   D. At the highest time of the process.

3. What is the role of a Kanban ?
   A. Increase value-added time.
   B. Accelerate the physical flow.
   C. Regulate upstream production according to downstream consumption.
   D. Reduce Takt Time.

4. What are the advantages of the VSM method ? (several answers possible)
   A. To visually represent a process.
   B. Validate the capability of a process in relation to customer demand.
   C. Identify the blocking points of a process.
   D. To be a strategic decision-making tool.

5. How are electronic information flows symbolized ?
   A. By dotted arrows.
   B. By continuous arrows.
   C. By lightning arrows.
   D. By FIFO arrows.

Now let's look at the following physical flow:

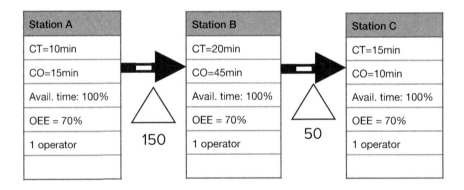

6. What's the bottleneck position in the process :
   A. Station A.
   B. Station B.
   C. Station C.

7. Stations A, B and C are open for 7.30 h, including a 30-minute break. The customer removes 25 pieces per day. What is the Takt Time ?
   A. 16,8 minutes.
   B. 12,3 minutes.
   C. 20,8 minutes.
   D. 20 minutes.

8. What is the process lead time ?
   A. 2 days.
   B. 6 days.
   C. 8 days.
   D. Some information is missing to answer !

9. What is the point of setting up a continuous flow ?
   A. Reduce work in process.
   B. Achieve customer Takt Time.
   C. Increase inventory.
   D. Improve the reactivity rate.

10. The production order will generally be injected (several answers possible) :
   A. In the last step of the process if the previous steps are linked by a pull flow.
   B. At the first process step if the subsequent steps are linked by a pull flow.
   C. At the last process step if the previous steps are managed in FIFO or continuous flow.
   D. To the first process step if the subsequent steps are managed as FIFO or continuous flow.

Answers : 1.A / 2.C / 3.C / 4.A, B, C and D ! / 5.C / 6.B / 7.A / 8.D (upstream and downstream stock information is missing !) / 9.A, B and D / 10. A and D.

NOTES

NOTES

Printed in Great Britain
by Amazon